WÄW

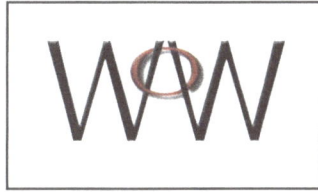

Choosing a Career in the Pulp and Paper Industry

Logs being transported down river to the paper mill

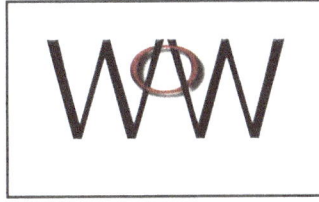

Choosing a Career in the Pulp and Paper Industry

Allison Stark Draper

The Rosen Publishing Group, Inc.
New York

Published in 2001 by The Rosen Publishing Group, Inc.
29 East 21st Street, New York, NY 10010

Copyright © 2001 by The Rosen Publishing Group, Inc.

First Edition

Library of Congress Cataloging-in-Publication Data

Draper, Allison Stark.
  Choosing a career in the pulp and paper industry / by Allison Draper. — 1st ed.
        p. cm.—(The world of work)
  Includes bibliographical references and index.
  ISBN 978-1-4358-8692-6
     1. Paper industry—Vocational guidance—Juvenile literature.
[1. Paper industry—Vocational guidance. 2. Vocational guidance.] I. Title. II. World of work (New York, N.Y.).
HD9820.5 .D73 2000
676' .023'73—dc21
                                                                00-009756

Manufactured in the United States of America

Contents

1

From Wood to Paper

Wood may be the oldest building material on the planet. Long before sawmills started turning trees into planks, people were tying animal hides over saplings to make tents and tepees or stacking logs into walls to make log cabins. Today, growing and harvesting the trees that people use to make buildings and books is the job of the pulp and paper industry. The companies in this industry own land all over the country. They own forests in most of the states in the nation. There are ten million acres in Maine alone that are paper company land. Many companies own forests in other countries, too, such as Canada, Finland, and Brazil. Some mill the trees that turn into books, magazines, liner notes for CDs, and napkins. Others cut trees for plywood, packing crates, and two-by-fours. Every plank in a house or leg on a chair comes from a tree in a forest that is probably owned by a paper company.

Wood is measured in cords. Stacked in a lumberyard, a cord is four feet wide by eight feet long by four feet high. A cord of wood can make 30 rocking chairs, 7,500,000 toothpicks, 1,000 to

2,000 pounds of paper, 61,370 standard-sized envelopes, or 4,384,000 commemorative postage stamps. Wood can also be chipped into tiny pieces, mixed with water or chemicals, and mashed into pulp. Wood pulp is used in rayon, laundry detergent, camera film, tires, and transmission belts. Most importantly, it is used to make paper.

Paper is one of the most important materials in the world. It is the basis for books, magazines, newspapers, letters, envelopes, cardboard boxes, and paper bags. Until e-mail, it was the medium of all correspondence. It is still the final surface for charts, graphs, memos, reports, and term papers. Many jobs would be impossible without the use of a lot of paper; doctors write prescriptions, lawyers write briefs, surveyors draw maps, teachers write tests and assignments, students write papers, politicians write speeches. Hosts send invitations; guests eat off paper plates, use paper napkins, and send paper thank-you notes.

In addition to writing paper—bond paper and letterhead and stationery and the crisp white sheets of $8\frac{1}{2}$ x 11 computer paper that come in those big cardboard boxes that are so terrific for storing things—there is paper packaging for food, clothing, and furniture. Every store, every super-market, every delivery company houses and uses thousands of boxes every day. Some are strong enough to support the weight of a person. Stronger still is construction paper. Construction paper includes real building materials made of paper and used to build and insulate houses and warehouses. According to the Institute of Scrap Recycling

Wood is chipped into tiny pieces, mixed with water or chemicals, and mashed into pulp to make paper.

Industries, more than one and a half million tons of paper construction products are made every year. These include insulation, gypsum wallboard, roofing paper, flooring, padding, and sound-absorbing materials. Strong, durable paper is one of the most common building materials in the world—just like wood.

The steps for turning hard wood into paper or cardboard involve a lot of in-between stages. To pulp something is to beat it into mush. In the papermaking industry, pulp is the word for wood that has been cut into tiny pieces, mixed with water, and mashed into a kind of paste. When this paste is smooth and even, the water can be pressed out of it, leaving behind paper or cardboard. The process for making paper is straightforward, but the idea is not obvious. It took a lot of time and a lot of clever people to come up with the idea for making paper out of wood.

2

The History of Paper

Wood was not always the raw material for making paper. The first paperlike writing material was probably Egyptian papyrus. Papyrus is a long, flat reed that grows in marshy areas or on pond edges. It can be woven and pounded into flat sheets that provide a good surface for writing or painting. The fact that it grows only in Egypt and Sicily, however, made it impractical for most countries. People who could read and write usually painted or embroidered onto fabric, or carved their words into stone. Real paper as we know it today was not invented until around 105 AD. This paper was made from cotton and linen rags by a Chinese court official named T'sai Lun. T'sai Lun's ingredients were probably mulberry bark, hemp, and rags. He combined them, added water to make a soupy mixture, and then mashed the mixture into a smooth pulp. This was comparable to modern wood pulp. T'sai Lun then pressed the liquid out of the mixture. This left a thin mat of "paper." Dried in the sun, the mat crisped into a flat page smooth enough for writing or drawing.

The first paperlike writing material was probably made from Egyptian papyrus, a long, flat reed that grows in marshy areas.

The Chinese kept the papermaking process a secret for many years. Other cultures continued to use stone and fabric. Slowly, however, people began to invent their own versions of T'sai Lun's cloth paper. More people learned to read. This meant that the demand for paper began to rise. Unfortunately, cloth paper was extremely expensive. (Even today, very expensive papers—like financial documents, stationery, and paper money—have a high rag content, because cotton fiber makes paper that feels nicer, heavier, and more flexible than wood.) The expense of paper was not a problem at the Chinese Imperial Court or in the palaces of Europe, but it quickly became one as more people began to use the written word. In the Middle Ages, anyone who could read was usually very rich or part of a religious order (priests, monks, and nuns were often highly educated, whereas shopkeepers, who did immense amounts of math in their heads or by making marks on wood, could write only their names). By 1690, when William Ritten House and William Bradford were collecting old cloth rags to make America's first writing papers at Wissahickon Creek near Philadelphia, Pennsylvania, it had just become common for ordinary people to read.

Less than one hundred years later, America's first official paper merchant was Benjamin Franklin. Franklin helped to start eighteen paper mills in and around Virginia. The Americans were fighting the British in the Revolutionary War, and paper had acquired another important use—rifle wadding. Seventeenth-century rifles needed to be

America's first official paper merchant was Benjamin Franklin, who started eighteen paper mills in and around Virginia.

loaded for every shot. A hunter or soldier poured gunshot and gun powder (which was also invented by the Chinese) into the barrel of a rifle and then pushed in a wad of paper, packing it down with a long thin metal ramrod. The paper held the shot and powder in place until the explosion of firing. During the American Revolution, one of the supplies that ran low among the American soldiers was paper. They were forced to tear pages from books in order to have enough wadding to keep up the attack on the British.

A war is an unusual situation, but it was clear that paper was so useful and so necessary that someone had to find a cheaper way to manufacture it. In the early 1800s, a Frenchman named Nicholas-Louis Robert invented a machine that would turn pulp into paper in a continuous roll. The machine was divided into four sections. The first part, called the "wet end," took pulp in through a wire screen. Then the "press" pressed out the liquid. The "drier" dried the flat sheets, and the "calender" smoothed and rolled the sheets into rolls of paper. This machine made it much faster and easier to make paper. It was soon patented by the Fourdrinier brothers in England.

Nonetheless, paper was still something of a luxury, and people are always interested in doing things more cheaply. About fifty years after Robert's invention, another Frenchman, a naturalist named Réné de Réamur, observed that the material of a wasp's nest is very similar to paper. Wasps, obviously, do not have a lot of access to rags, so what was this material? De Réamur became interested. He began trying to copy what the wasps had done using only the materials they could have used. His experiments interested other researchers, and their efforts eventually led to the making of paper from wood fiber. (This was an important discovery not only for paper technology, but because cheap paper helped to encourage more widespread literacy.)

In 1866, in the United States—which would not begin making paper from wood fiber until the early 1900s—an American named Benjamin Tilghman developed the "sulfite" pulping process. This was another huge step forward. In fact, sulfite pulping is still used in the papermaking process today. The first mill to use Tilghman's process was built in 1874 in Sweden, where much of the pulp and paper industry is still located. By 1937, the sulfite process was the major pulping process in the world.

Sulfite pulping is not the only pulping method. In 1879, a German chemist named C. F. Dahl developed the kraft or "sulfate" pulping process. Kraft is the German word for "strong." The kraft process had important advantages. First, the chemicals it used to dissolve the lignin—the natural glue that holds wood fibers together—were recoverable. (At that time, people were not worried about letting chemicals into the environment; they simply wanted to recover the chemicals in order to use them again and save money.) In addition, the process could pulp pine trees. The United States was covered with pine forest. The kraft method meant that the United States could harvest its pine forests. This is how America became a major international producer of paper products. The first kraft mill in the United States was built in 1911 in Pensacola, Florida. The kraft pulping process is still the major chemical pulping process in use today.

Since the early days of industrial paper-making, the steps of the process have not changed very much. What has changed is how quickly and cheaply paper can be made. The other major modern advance is the use of wood pulp in such non-paper products as laundry detergent and synthetic fabrics.

3

Modern Pulp and Papermaking

Today's papermaking machines are a lot like the early papermaking machines of the 1800s, but they are much larger and much faster. A modern papermaking machine can run at 2,500 feet per minute and make more than a million square feet of paper in an hour. It is fed by a pulping vat 30 feet wide and 20 feet deep that can hold 17,000 gallons—or 5,600 pounds—of pulp.

Modern papermakers make paper with fiber from both new wood and recovered paper. About 40 percent of today's paper fiber comes from newly cut trees. Leftovers account for 24 percent. Leftovers are scraps of wood like log ends that are left around after a log is cut up for lumber. They can be tossed into a chipper and ground up for pulp. The rest of the paper fiber in the United States comes from recycled paper. Papermakers are able to use so much recycled paper because Americans do a lot of their own recycling. More than 40 percent of all the paper used in America— as newspapers, orange juice cartons, and wrapping paper, among other things—is tossed

More than 40 percent of all the paper used in America is recycled. Paper trash can be recycled into new, strong, and useful paper products.

into green and blue plastic recycling bins in millions of homes and offices. This paper trash can be recycled into strong, useful paper.

The Papermaking Process

The first step in making paper is to make a pulp mixture. The point of pulping is to weaken the lignin. Lignin is the natural glue in wood that holds the wood fibers together. There are three different pulping processes: mechanical pulping (mashing by hand or machine), chemical pulping (breaking down the wood fiber with chemicals), and recycled pulping (there is no lignin in recycled paper, so the process is basically just paper shredding).

In all of these processes, wood chips and water are added to a big vat called a pulper. The mixture is about 1 percent wood fiber and 99 percent water. The pulper mashes or chemically treats the wood fiber. The fibers break apart and blend with the water. This makes a soupy mixture called slurry. For certain types of paper, papermakers may also add dyes or fillers to the slurry. Dyes give the pulp color and fillers make it heavier.

There are two major methods of mechanical pulping. The first is the stone groundwood process. In this method, electrically powered stones grind the wood fibers. The second process is called thermo-mechanical pulping, or TMP. TMP is different from groundwood pulping because it uses steam to soften the wood chips in a large vessel that looks like a giant pressure cooker. Heat weakens the lignin. This makes it easier to separate the fibers mechanically.

Chemical pulping uses the sulfate, or kraft, process. This process cooks the wood chips in a combination of chemicals in a high-pressure vat. The heat and pressure force the chemicals into the wood. The chemicals dissolve the lignin. The advantage of chemical pulping is that the pulp it makes can be used for more kinds of paper than mechanical pulp. Mechanical pulp has short fibers. It is used mostly for newsprint and printing and writing paper. Chemical pulp has long fibers. It is stronger, more durable, and has a better color than mechanical pulp. The downside is that chemical pulping uses only about 45 to 47 percent of the wood. Mechanical pulping uses 85 to 90 percent of the wood.

After it has gone through one of these pulping processes, the pulp is forced through a screen. This makes it smoother. Then it is cleaned and bleached to brighten it. Special kinds of paper, like fine art paper or glossy magazine paper, are usually made from several different pulps. Short-fibered pulps make paper that is soft, smooth, and opaque (which means you cannot see through it). However, short fibers are not strong enough to last on the high-speed printing presses that make books and magazines. For this reason, good printing papers are usually a combination of short-fibered mechanical pulps and long-fibered chemical pulps.

Once the pulp is ready, the papermaking process can begin. First the wet pulp is fed onto a wire screen at the papermaking machine's wet end. The screen rotates at high speed like a conveyor belt. As the screen moves, water starts

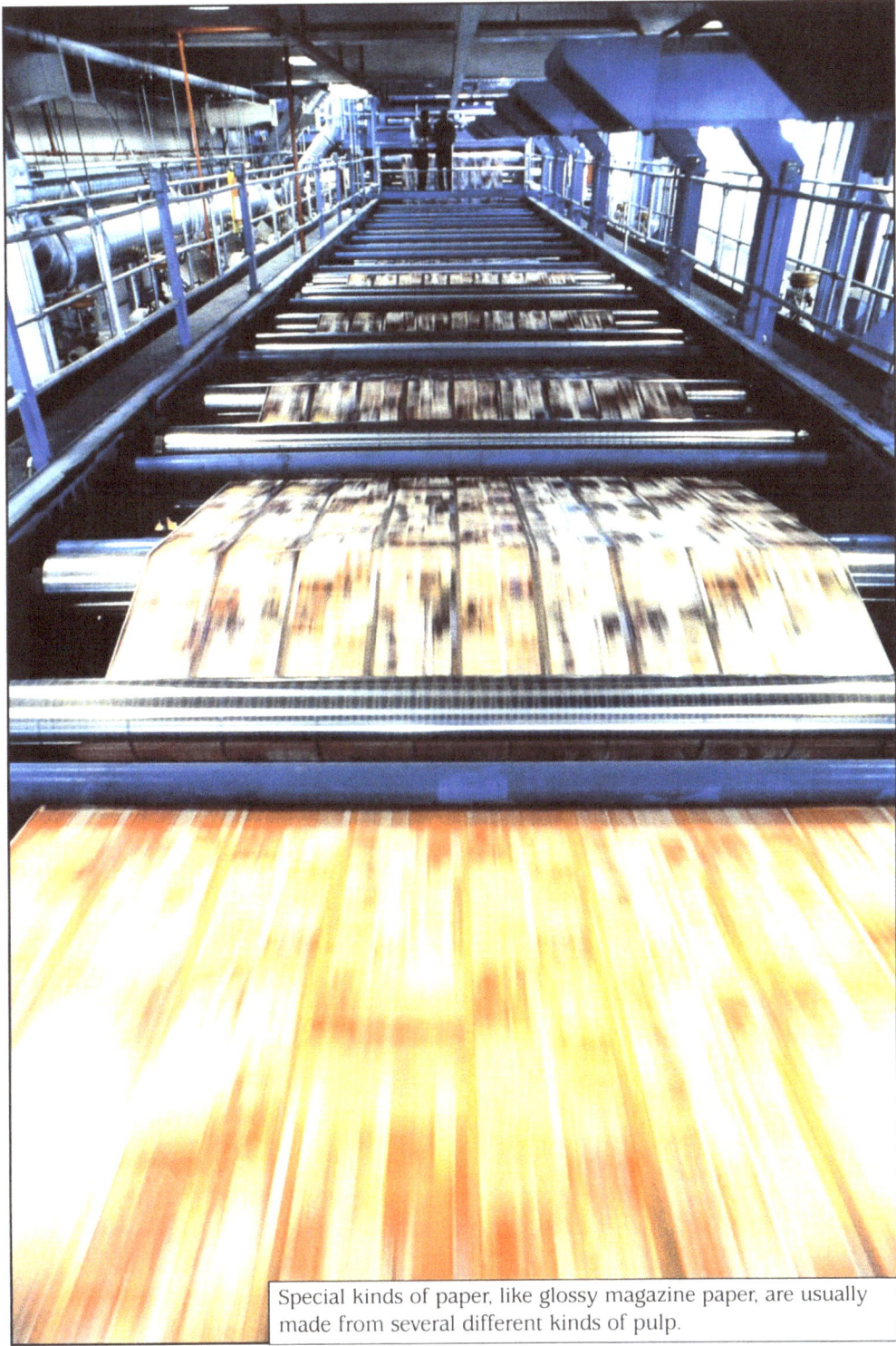

Special kinds of paper, like glossy magazine paper, are usually made from several different kinds of pulp.

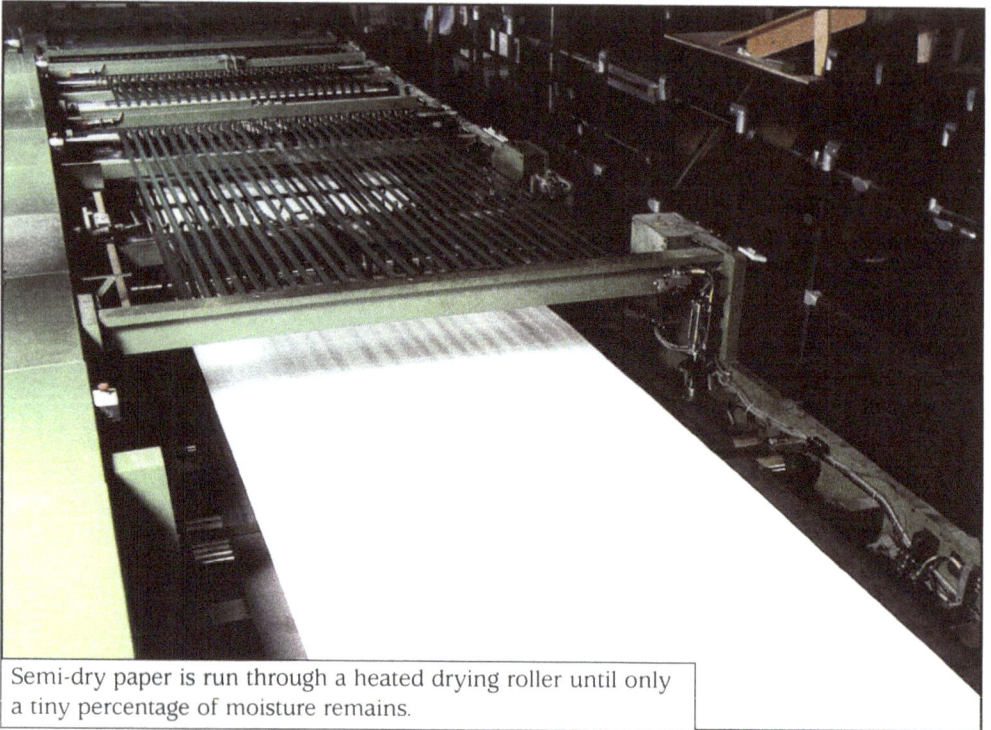
Semi-dry paper is run through a heated drying roller until only a tiny percentage of moisture remains.

draining from the pulp through the screen. The screen is now covered with a long ribbon of pulp. This wet, fibrous mat is what will become the sheet of paper. Next, the mat is squeezed between large rollers. This removes most of the remaining water and makes the mat smooth and even. The semi-dry paper now runs through a heated drying roller to remove the remaining water. It is rolled and dried until only a tiny percentage of moisture remains.

When the paper is ready, it is coated and treated for use. For instance, the glossy paper used in magazines and catalogs is coated on both sides with a mix of clay and starch. This gives it a smooth, shiny surface. After it has been treated, the paper is rolled. The largest papermaking machines have rollers around thirty-two feet wide. They are as long as 550 feet and weigh thousands of tons. Some can produce more than 1,000 miles of paper in a day. Once the paper has been rolled, it is rerolled onto a second set of rollers. As the paper passes from the first to the second set, workers check it for defects. They check the color, the smoothness, the water resistance, and the "inkability," which means how well paper takes printing ink.

When the paper rolls are approved, they are fed into supercalender machines. These machines have ultra-high-pressure rollers and smooth the paper even further. When the paper is finished, it is wound into parent rolls. Parent rolls can be thirty feet wide and weigh twenty-five tons. A machine called a slitter cuts the parent rolls into more manageable sizes. Some of these smaller rolls are cut and rewound for particular customers. Others are stored for later sale. An average "manageable" roll weighs between 1,000 and 7,000 pounds.

Pulp that is not sent directly to a paper machine is either sold for manufacturing or else dried and stored. The last step is to treat the leftover chemicals and waste water from the papermaking process. Pulping chemicals are very dangerous for the environment. Brown wood chips must be bleached to make fine white paper. Since much of the recycled paper used has printing on it, recycled slurry is often gray from old ink. Gray slurry also needs to be whitened with bleach. The bleach is a chlorine-based chemical that can be toxic. Paper manufacturers go to great lengths to remove such chemicals from the water their factories discharge.

After paper is finished being smoothed, it is wound into very large "parent" rolls.

Recycled Paper

Paper for recycling is tied into huge bales. These bales are forklifted and dropped into a hydra-pulper. A long rope called a ragger dips into the hydrapulper to fish out contaminants. These generally include metal bits like wire and staples, plastic tape, and occasionally pieces of clothing or jewelry. The hydrapulper then mixes water in with the recyclable paper. A large-bladed propeller at the bottom of the hydrapulper rotates at high speed. It shreds the paper into fibers. The lack of lignin means that the recycling process is purely mechanical. The recycling mill does not use any chemicals at all. This also means that unlike a paper mill, which usually smells terrible, a recycling mill just smells like wet cardboard.

The hydrapulper spits out any large non-fiber material and then sends the finished pulp to the pulping screen. The screen removes smaller contaminants. The pulp travels to the forward cyclones. It is whipped around until any heavy, tiny contaminants like chips of stone or glass are pulled to the bottom by centrifugal force. It then goes into the reverse cyclones, in which any light, tiny contaminants like bits of plastic are pulled to the top. In recycling, cleaning the pulp is very, very important. Unlike wood pulp, which comes from clean trees, recycled pulp comes from garbage. After the cyclones, the newly cleaned pulp goes to the refiner, which "brushes" the paper fibers to thicken them. Thicker fibers make stronger sheets of paper.

Next, the pulp is divided. Some will be used to make the bottom ply of the paper. This part goes straight to the headbox of the papermaking machine. The rest of the pulp will become the top ply. The top-ply pulp passes through another set of cyclones so that it can be cleaned again before it is sent to the headbox.

The headbox is a box with nozzles. When the pulp is poured into it, the headbox sprays the pulp out through its nozzles onto a moving screen. This is the wet end of the paper machine and is the same as in papermaking from wood pulp. The pulp moves down the screen toward the press section, the water drains, and the pulp begins to form a sheet of paper. Felt conveyor belts carry the paper through the press section where the paper is still 40 percent fiber and 60 percent water. The next set of felts takes it to the dryer section. Steam-heated rolls dry the paper. Now the paper is only 7 percent water. At the end of the dryer section, calender rolls polish the paper to a smooth finish.

The recycled paper is now ready for use—without the death of a single tree.

4

The Tree

The resource for the pulp and paper industry is the tree. To make sure that there are enough trees for all the stationery, picture books, newspapers, and napkins the world needs, the paper industry grows trees and harvests them like a farm crop. Pulp and paper companies manage huge forests the way farmers manage corn and wheat fields.

Tree farming is unusual because, unlike corn or tomatoes which grow from seed to product in a single season, trees need years to reach maturity. Different kinds of trees grow at different speeds and fast-growing trees are not necessarily the largest. Trees also live for different lengths of time, depending upon their species, climate, and situation. Some trees, if untouched by storms, fire, disease, or humans, will live for thousands of years. The oldest living thing on the planet is very probably a yew tree.

A thousand years is a long time to wait, but many companies in the pulp and paper industry are increasingly aware that keeping some trees of

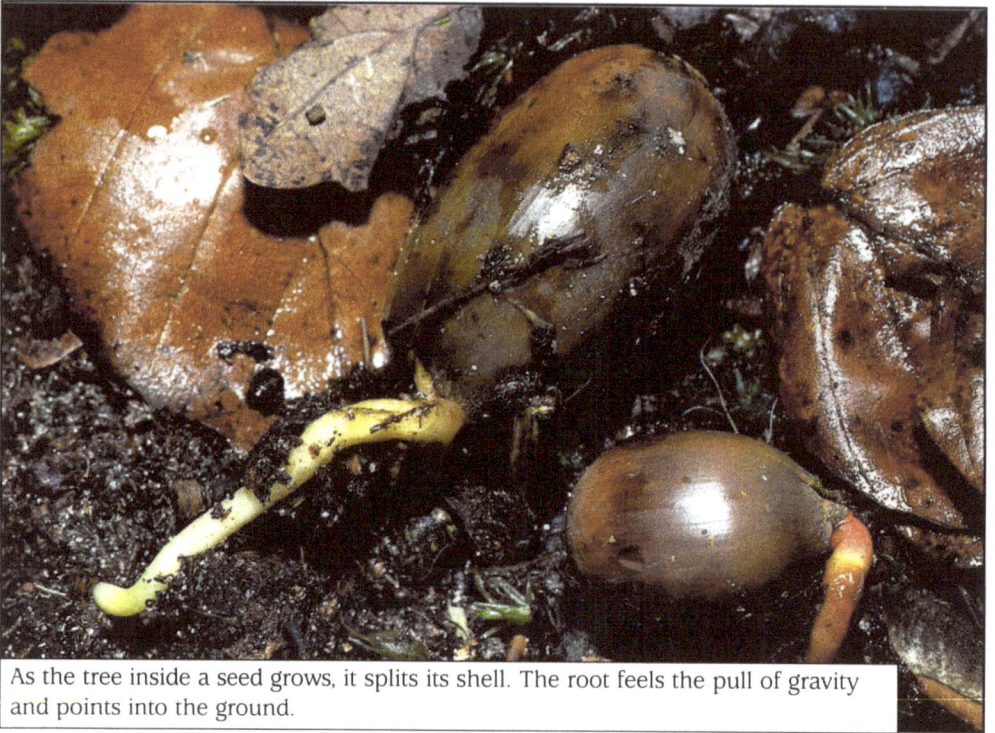

As the tree inside a seed grows, it splits its shell. The root feels the pull of gravity and points into the ground.

every kind and size and age alive is a good thing for the health of a forest and the health of the planet. This is called biodiversity. A forest that provides homes for many different plants and animals and insects may include animals that eat insects that feed on tree leaves. Certain types of birds may spread the tree's seeds, allowing new trees to be created far from the original tree.

A tree seed contains a tiny baby tree complete with leaves, a stem, and a point that will become a root. As the tree inside the seed grows, it splits the seed shell. Its root feels the pull of gravity and points into the ground. When the root enters the soil, it begins absorbing the water and nutrients the tree needs to grow larger. The leaves emerge from the shell and fan out to absorb sunlight. The green material in the leaves is called chlorophyll. A tree uses chlorophyll to make food from sunlight. At the center of the leaves is a tiny terminal bud. All the upward growth of the tree starts from this bud.

As the tree grows up, it also grows out. A thin layer of living cells surrounds the wood of the tree's trunk just beneath the bark. It creates new wood on the inside, and it creates new bark on the outside as the tree grows bigger. It increases the width of the tree as it moves outward, growing more bark around a broader core of new wood. The outer bark protects the tree by making sure that the right amount of moisture reaches the inside wood. It keeps water out during rainstorms and holds it in during droughts. The bark also guards the tree from sudden changes in

A tree's upward growth toward sunlight starts from the tiny terminal bud located at the center of the leaves.

temperature, and its hard, scaly surface stops insects from burrowing into the tree's wood.

The tender lining of the bark is called phloem. This lining carries food from the leaves to the rest of the tree. The phloem is constantly renewed. The cells of the phloem layer live for a very short time before they die and turn to cork. The cork then becomes part of the protective outer bark.

The growing interior of the trunk is the cambium layer. Every spring, when the leaf buds at the tips of the branches begin to grow, they produce hormones called auxins. The auxins travel through the phloem along with food. They tell the cambium layer to produce new bark and new wood. New wood grows in the spring and summer. In the spring it grows faster and is made up of larger cells. In the summer, it grows more slowly. When a tree is cut, this change in speed from spring to summer growth appears as lighter (spring) and darker (summer) rings. If you count either the light or the dark rings, you can tell the tree's age. In addition, the rings reveal the events of the tree's life. Full sunlight, shade from competing trees, fire, drought, and insects can all change the shape, thickness, color, and evenness of the rings.

New wood is called sapwood. Sapwood is the soft, outer ring of wood that carries water from the roots to the leaves. As new sapwood grows, the older layers of sapwood dry out and become heartwood. Heartwood is what supports the whole, enormous weight of a tree. It is a pillar of now-dead wood inside the still-growing layers of the tree's outer wood and bark. As long as the outer layers of a tree

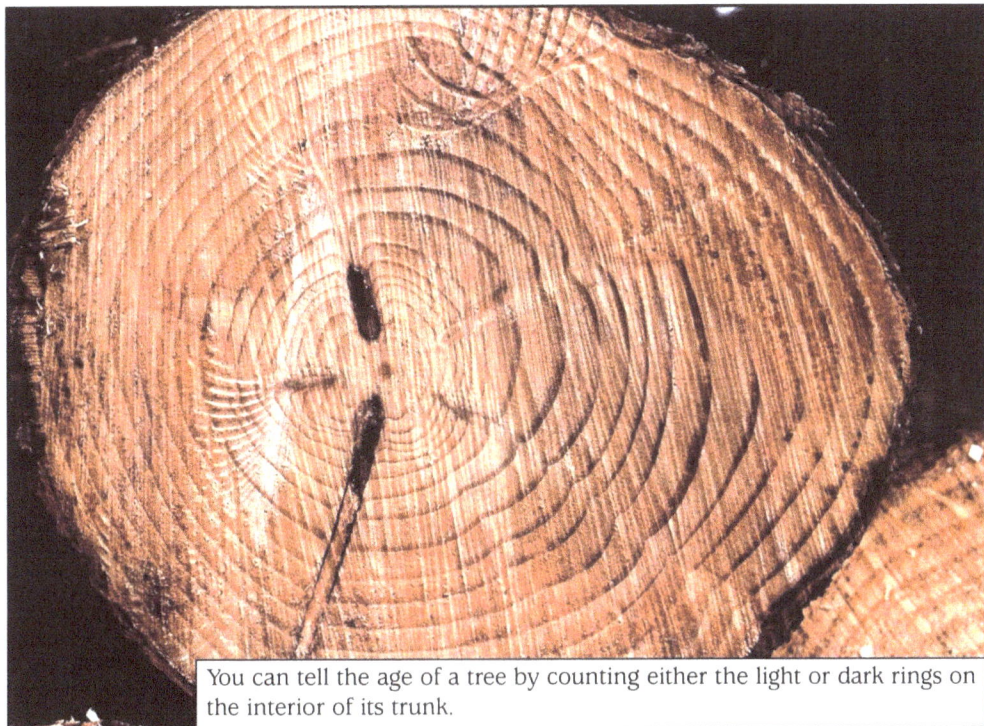
You can tell the age of a tree by counting either the light or dark rings on the interior of its trunk.

are healthy, the heartwood will not weaken or decay. Depending upon how it is used, heartwood can be as strong as steel. A twelve-inch by two-inch by one-inch piece of heartwood, set vertically, can support twenty tons. Heartwood is so strong in part because of lignin. Lignin is the glue that binds together the fibers in a piece of wood. One of the main jobs of industrial wood pulp processing is to remove the lignin.

The tree is a simple but highly effective design. A tree may live for centuries, and it builds a trunk that may eventually support a weight measured in tons.

5

The History of American Logging

Filled with strong, ancient trees, the old-growth forests of North America were an obvious source of wealth to European settlers. The North American logging industry has existed since the 1600s, when French and British colonists settled the forests of New England and Canada. Until loggers started to use trucks in the early twentieth century, logs traveled in wagons pulled by horses and oxen, by railroad, or on water. The largest logs are the most valuable, but large logs are hard to transport on land. In the water, however, they float. Early logging tended to be limited to areas near waterways. Logs traveled directly from the stump to the mill.

In the nineteenth century, loggers clearcut the white pine forests of the northeastern and lake states. They tied logs together into huge rafts and lived on them. They ate and slept in floating bunks called wanigans. When they had as much wood as they could manage, they floated the rafts out of the back woods to the riverside milltowns where they could be sawed into planks or pulped into paper. In the spring, rain and snowmelt raised rivers to

Nineteenth-century loggers used white pine logs to make rafts and bunks.

flood level, making them wider and faster. Loggers rushed to transport huge shipments of logs to the mills. Flooded rivers are dangerous. Sometimes colossal log jams occurred, cracking up the loggers' huts and trapping them among angry tons of wood. Sometimes, toward the end of the flood season, the water would drop suddenly. Loggers would find huge logs stranded out of reach of the river. They would have to drag them back down into the shrunken streams.

Logging was—and is—back-breaking work. In the early days, loggers worked with axes and hand saws. Some trees were so large that two loggers had to work together, one at each end of a two-handed saw, to cut them down. In the old forests of the East, trees were large. In the West, they were immense. Trees with trunks six or seven feet across were not unusual in Colorado and California. In

Logging is very difficult work, especially when cutting the very large trees of the American Northwest.

addition, the West had few rivers and fewer lakes. In the pre-trucking days, the western loggers had to build flumes to transport logs. A flume is a water channel built out of wood. Logging flumes had some advantages over rivers. They were designed to run straight and to fit one log. The boulders, waterfalls, and flooding problems that made rivers so dangerous did not exist. On the other hand, flumes were expensive to build. Every stick that went into building a flume was one less piece of wood for the logging companies to sell. This meant that loggers used up large amounts of wood in order to transport and sell less lumber.

The logging industry wanted a better method for moving logs. After experimenting with steam power and railroads, loggers discovered the tractor, an ideal machine for plowing out roads through the forest. The tractor traveled on wheels with rotating treads. Loggers added blades to the fronts of these tractors to cut out low scrub and old trunks. Wheeled vehicles could now drive into parts of the forest. This meant that loggers could carve out stretches of dirt to act as firebreaks in case of forest fires. By digging out all tree and plant material from a piece of land, loggers were able to make the land unburnable. This meant that any forest fire could be contained inside a boundary described by bulldozer blades. Loggers began constructing these firebreaks as a matter of course. Without knowing it, these loggers were reshaping the ecosystem.

A natural forest grows as the direct result of what has died in it. This is called stand renewal, because a group of trees is called a stand. In an unharvested

Today, trees are removed from forests by truck on roads created by tractors.

forest, stand renewal happens constantly. New seeds take root while some trees are saplings, others are in middle age, and others are dying and decaying. The shape of a forest changes from year to year, depending on which seeds take root, whether it is hot, cold, or wet, and whether there are violent wind or lightning storms. In addition, stand renewal often involves wildfires. Forest fires are destructive, but they can also be good. They enrich soil by speeding decay. Decay releases nutrients that cause stronger new growth.

In the early part of the twentieth century, the American logging industry did not know or worry about any of this. The forests in the United States seemed vast and endless and full of perfect trees. Scientists had done little research about conservation or overharvesting. The loggers did not understand the importance of maintaining the natural condition of the forest. No one even guessed there was value in forest fires.

The tractor had other ecological effects. It created logging roads that reached diagonally up across hillsides. This meant that loggers could reach the highest, oldest stands of timber, which prior generations of loggers had left alone. Much old-growth woodland was lost. The tractor allowed loggers to clear a field completely and reseed it with only one type of tree. This created monoculture or single-species fields—the exact opposite of biodiversity. Tractor treads churned the forest soil with a tilling action that made it ideal for certain pine trees. This made money for the companies (pine is easy to sell) but was bad for the health of the forest.

Unlike a farm crop, which reaches maturity in a single season, trees need several human lifetimes to reach their full growth. During this time, other trees live and die all around them, offering different amounts of competition and different habitats for animals or insects that may be destructive or beneficial for all the other trees. This level of biodiversity makes a forest more fertile and more resistant to destruction. A forest of only one type of tree could be entirely destroyed by one insect in one season.

This was less well understood one hundred years ago—when the demand for pine caused loggers to leave spruce and fir rotting in the woods and to replant many acres with pine trees. For human beings, the life of a forest is a slow one. Someone who knows how to look will see that the forests of North America are still visibly recovering from the early years of logging and the introduction of the tractor.

6

Harvesting Trees
for the Sawmill

Most modern pulp and paper companies are serious about maintaining the health of their forests. The future of their business depends upon it. They grow and harvest trees for papermaking like a crop, but they maintain different species, nurture and protect areas of biodiversity, and plant millions of new seedlings every day.

The modern process of raising trees is very sophisticated. A new generation of trees is started indoors in a greenhouse or nursery. The seeds are planted in raised beds in nutrient-rich soil. This gives them the strongest start possible. For the first year, they are carefully watched, watered, and fertilized. As yearlings, they have sturdy root systems and full foliage. Now they can be planted in a prepared field from which a grown stand of trees has been cut. They are planted in rows with enough space to give each tree enough light and room to grow.

A team of ten loggers can plant 7,000 seedlings a day or 560,000 trees in a season. Seven thousand trees planted eight feet from each other

A seed orchard is a section of a tree nursery where trees with certain qualities are grown from cuttings for seed.

cover about an acre of land. When the seedlings are ready to be planted, the nursery packages them in bundles of fifty. The planters carry them in watertight tree bags the size of shopping bags that they wear on their hips. Each bag holds three or four bundles and weighs thirty or forty pounds. The planting tool is called a hoedag. It has a wooden handle and a long, flat, rectangular head. A planter swings the hoedag sideways to clear a square foot of land and then uses its short, sharp end to punch and widen a central hole of ten to twelve inches. The roots of the seedling are inserted into the hole and the earth is tramped firmly around them.

Each year, the new trees produce seeds that are extracted and kept in cool, dry storerooms to be used to grow future generations of trees. In addition, cuttings are taken from these trees and grafted to the stems of young rooted trees in a seed orchard—a section of a nursery where trees with certain qualities are grown for seed. The cutting takes food and water from the tree to which it is grafted, but it does not grow into the tree. In fact, a cutting may be grafted to a tree of a different species.

When the grafted cuttings bear flowers, the flowers are covered with plastic bags. This prevents them from being accidentally fertilized by strange pollen. Instead, the growers inject the bag with pollen from another tree in the seed orchard. Choosing pollen this way lets growers breed trees with certain qualities, like straightness, pest resistance, or speed of growth. A grower might cross a disease-resistant cutting with pollen from a fast-growing tree and make a new tree that is both disease-resistant and fast growing.

When a healthy tree is tall enough to be harvested, it is felled, cut into lengths, and transported by truck to a paper mill's woodyard. All parts of the tree are used. The first step in processing any tree at the sawmill is passing it through a debarker. Bark cannot be used for paper, but it can be used for fuel and soil mulch. The next pieces to come off are the rounded sides of the log, called "slabs." The slabs go directly to the chipper, in which spinning blades cut the wood into one-inch pieces. Later, wood from the chipper will be pulped for paper.

Without its slabs, the log looks like a square column. In a lumber log, every part of this column is used for boards and planks. The outer portions of the square column of the log are best for planks because they are clearest of dark spots or knots. As a tree grows, its lower branches drop away and its new skin forms over the holes or knots where the limbs used to be. This clear lumber is usually made into thin, one to three-inch boards. The wood in the center of the log, although knottier and less attractive, is still solid and strong. It is used for heavier planks and for square or rectangular beams that are thick enough not to be weakened by knots.

Another use for the outside of a log is plywood. Plywood is a sandwich of thin sheets of wood made by peeling a log. A long blade slices a thin layer from a rotating log in a continuous sheet. The sheet is pressed with other sheets to make flat

When trees are tall enough to be harvested, they are cut down, chopped into shorter lengths, and transported by truck to a paper mill's woodyard.

rectangles of plywood. The plywood peeling process leaves an eight-inch core of wood that is treated like a small log. Its slabs go to the chipper and the rest is cut into planks. The wood chips in the chipper are mixed with water and mashed into pulp. The pulp is either used to make non-paper products or sent on to the paper machines.

7

Working in the Pulp and Paper Industry

The modern papermaking industry invests a lot of money in equipment and technology. Most papermaking companies spend more than $100,000 in equipment and support per employee. Inside a paper mill, the people most involved with the papermaking process are the technicians who run the papermaking machines and package the paper for delivery to customers. Overseers monitor the technicians. Managers oversee the overseers and handle accounting. Sales people deal with customers and orders. Marketing people tell the public about the company and its products.

Students who want to develop, produce, and market paper and wood products usually study wood and paper science. They explore the chemical, physical, and mechanical properties of wood. They study the technologies for turning wood into paper or furniture. After learning about the industry, a wood and paper science student will usually choose to be in marketing, production, mill operations, or research.

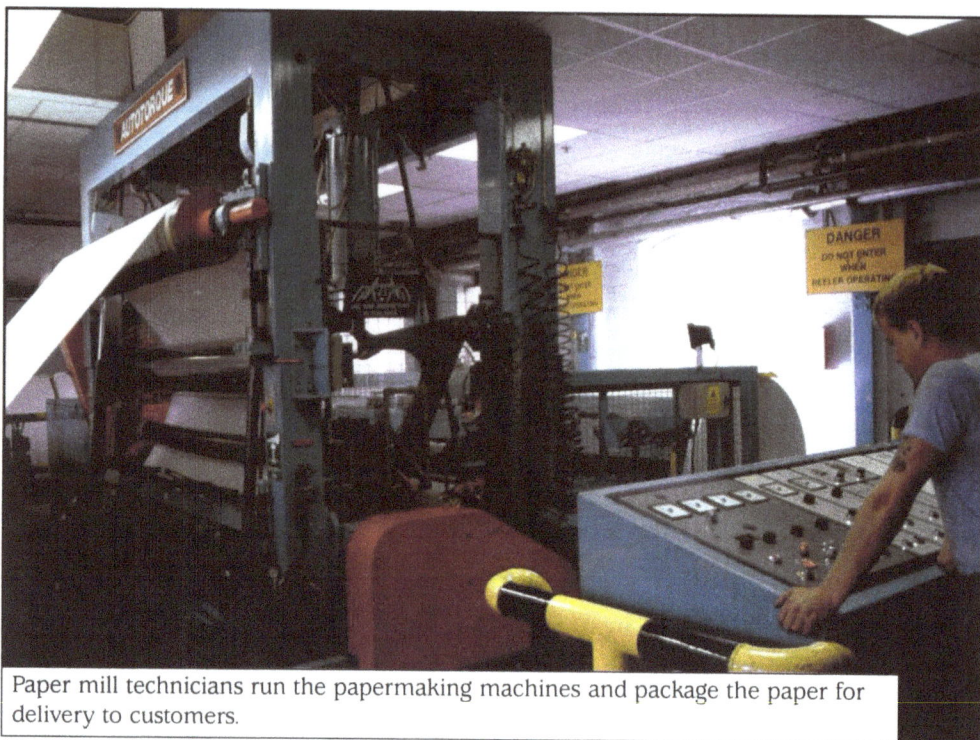

Paper mill technicians run the papermaking machines and package the paper for delivery to customers.

Marketing courses teach students about the sale of forest products like paper and particleboard. They explain the physical nature of building materials like lumber and plywood. A person with a degree in forest marketing can become a seller of forest products for a paper company or a buyer of forest products for a retail company, like a furniture maker or a shipyard.

Students interested in the actual processes of making wood and paper products can learn how to run and manage paper mills. These students research new wood products, like composite building materials. They study industrial engineering and economics. They find jobs working for companies that manufacture lumber, panel products, millwork, or furniture.

If a student wants to know more about the technology behind wood construction, there are areas of wood science programs that focus on these questions. They explore the energy efficiency of buildings built out of wood. They compare this to the energy efficiency of buildings made of non-wood or composite products. They look at how long wooden buildings last, how much maintenance they need, and how good their air quality is.

Students who pursue paper science engineering need to understand every step of the pulping and papermaking processes. They study biology, chemistry, physics, engineering, and manufacturing. This background prepares them for careers in paper engineering, plant management, corporate management, or research and development. Research is one of the most important parts of the pulp and paper industry. The idea of sustainable forestry is still quite new. There is much to learn about using and conserving forests. There are forestry researchers who study ways to help trees grow faster, straighter, healthier, and larger. There are papermaking researchers who invent methods for making paper whiter, smoother, and less expensive.

Currently, one research goal is to increase pulp yield. This means that researchers are trying to get more usable pulp from a single tree. Increased pulp yield would make paper mills more profitable, and—just as importantly—would mean that logging companies could cut fewer trees for the same amount of paper.

A big concern of the paper industry is whether wood is really the best material for making paper.

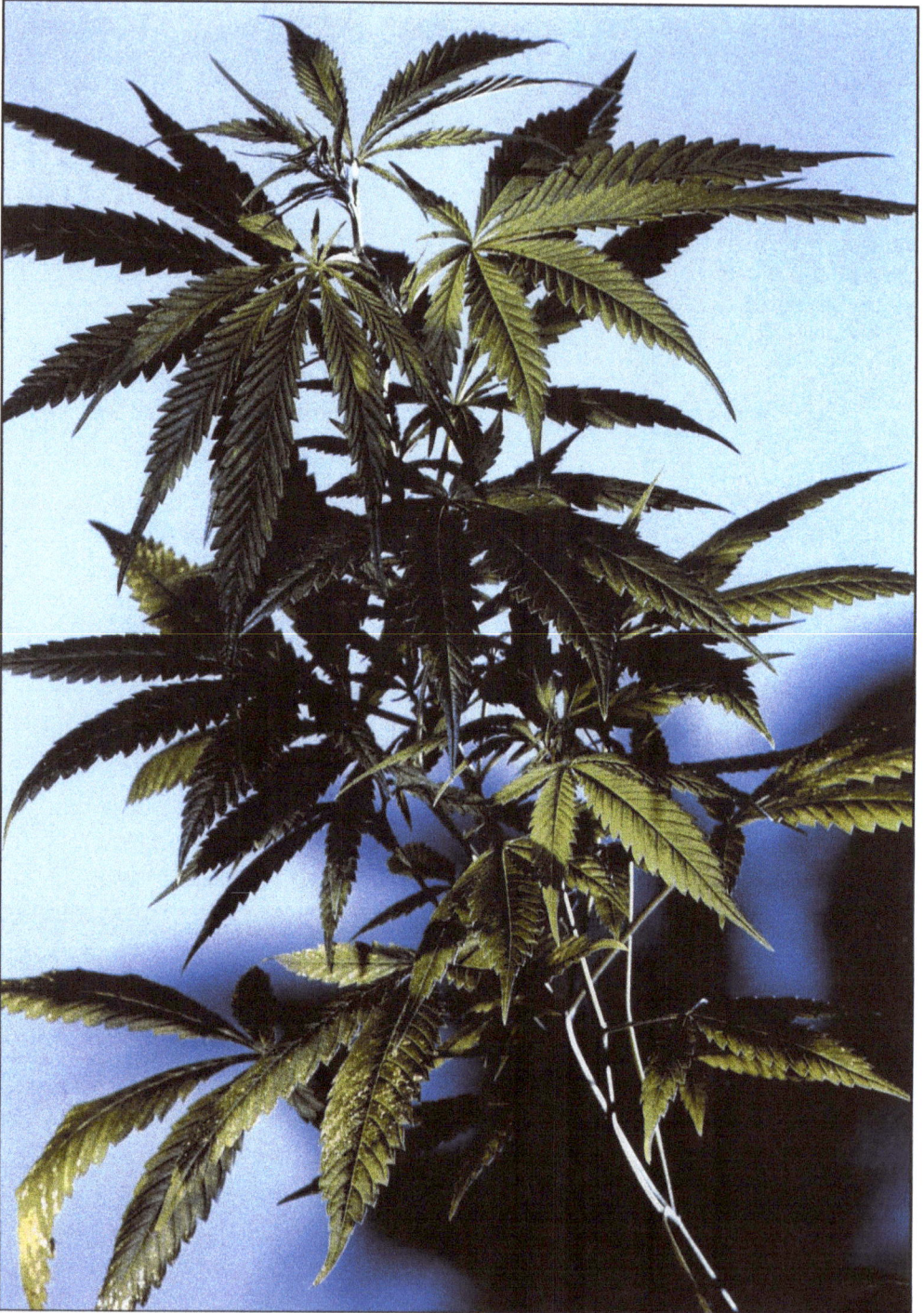

The hemp plant is a potential source of papermaking that is an alternative to trees.

About 15 percent of the wood taken from forests is used by the pulp and paper industry. Some people feel that another plant should be cultivated specifically for papermaking so that forests will not be overharvested. One possible plant is hemp. Researchers are studying the best time to harvest hemp for pulping. Another possible alternative is kenaf (*hibiscus canabinus*), which, like hemp, is a non-woody plant with pulpable fibers.

Like all the people in the pulp and paper industry, whether technicians, marketers, managers, or loggers, these researchers are seeking to improve their companies' products while preserving the health of their forests.

8

The Future of Pulp and Paper

When people today use the word sustainable, they are usually talking about taking care of the environment. The word was first used this way by the foresters of eighteenth and nineteenth-century Europe. At that time, Europe was being deforested. Foresters in heavily wooded countries like Germany were worried about their livelihood. They were also worried about the continuation of normal life. Wood was of great importance. Europeans used it to build houses, barns, wagons, carriages, and furniture. They burned it in fireplaces and wood stoves to heat their homes.

From the money standpoint, then as now, clearcutting made sense. Eighteenth- and nineteenth-century loggers marked off large tracts of land and cut them to the ground. The problem was that when the forests grew back, they did not necessarily provide as much wood. The foresters in Germany had an idea for how to solve this problem. They called it "scientific" or "sustainable" forestry. They decided that they would not simply wait for the forest to grow back. They would actively replant it. In this way, they would have control over the tree production of the forest. If they planted enough trees, they would never run out of wood. All they had to do was make sure that a few more were planted each year than were cut. They would also have to watch the forest to make sure that it grew back straight and healthy. They would have to maintain it like a farm, clearing dead trees that interfered with live ones and making sure that each tree had adequate water and light. The original idea of sustainable forestry was the same as today's idea. Foresters grew trees to replace the ones they took.

The question that faces the pulp and paper industry in America today is the same question that faced the German foresters. Will we run out of trees? The American need for paper and wood products rises by about 4 percent every year. This means that forest management has to be a little more productive and a little more efficient every year. Companies have to learn to use every millimeter of every log. And they are learning. At one time, 50 percent of a log was routinely thrown away. Now these waste pieces are thrown into chippers and used for paper. The millions of tons of chips in American paper have saved hundreds of thousands of acres of forest from being cut.

In the 1920s, people believed that Americans used so much paper and wood that they would use up the country's forests by 1945. The United States now actually has 20 percent more trees than it had in the 1970s. This does not necessarily mean it has more wood. A few large oaks represent a lot more

To protect forests, many different kinds of trees must be replanted to encourage biodiversity.

wood than a whole field of tiny pines. It does mean that logging companies are doing more replanting. Modern forestry has helped forest growth to significantly and steadily exceed harvest. The conviction of the pulp and paper industry is that forests are a viable renewable resource. The trick is to renew them more quickly. The modern pulp and paper industry tries to take just a little less than the forest can provide.

Protecting forests is not just about replanting trees. It is about replanting enough different kinds of trees to encourage biodiversity. In the same way that the government sets aside tracts of land for national parks, certain forestry companies set aside sanctuaries. They do no commercial cutting in these areas and encourage the growth of different plants and animals. This is good for the paper industry as well as for the environment. It maintains a pollen factory for the growth and cross-breeding of new and better trees. These sanctuaries are like living libraries of material for the ongoing health of the world's forests. Tree breeders can use pollen from the wild trees that self-seed in a sanctuary to increase the strengths of their nursery trees.

There are subjects on which foresters and environmentalists continue to disagree. The biggest environmental complaints about forestry are probably clearcutting (in which a forest is cut to the ground rather than thinned) and monoculture (in which a clear-cut area is replanted with a single type of tree). These practices are considered hazardous by many environmentalists because they endanger species and habitats.

Clearcutting, in which a forest is cut to the ground, and monoculture, in which a clear-cut area is replanted with one type of tree, are concerns of environmentalists.

People who object to the paper industry's cutting methods often want to try to find another plant to use for pulp. So far, there is no good alternative. Most possibilities are too expensive and some are worse for the environment than trees. Hemp, mentioned earlier, is fast-growing and non-woody. It produces useful fibers and oils. However, only 25 percent of it can be used to make paper.

Kenaf is related to cotton and okra. In Asia and Africa, people have used it to make clothing and twine for centuries. It likes warm, wet weather and can grow to be fourteen feet high in 180 days. Its pulp can be used to make newsprint, fiberboard, cat litter, and, possibly, car parts. But only 30 percent of it is useful for making writing and printing paper.

Alternatives to trees sound interesting, but so far none of them measures up to wood. Overharvesting could certainly destroy the world's forests. However, planting a lot of something else will not necessarily improve the environment. In addition, the paper industry is close to 100 percent efficiency in terms of using every fiber from the trees it cuts. It continually increases the amount of paper it recycles. Everyday, American papermakers recycle enough paper to fill a fifteen-mile long train of boxcars.

The paper industry has grown far more sensitive to the environment in recent years. If it continues on this path, the need to find another way to make paper may disappear. American forests need to be treated with intelligence and respect. The pulp and paper industry needs to keep its forests a little larger than the size it needs to harvest wood. This is sustainable forestry. Truly sustainable forestry could keep the forests safe for centuries to come. Look out your window; some of those trees may still be standing in the year 3000.

Glossary

auxins   Substances inside trees that make them grow.

biodiversity   The presence of many different plants and animals in one area.

calender   A very fast roller that smooths paper at the end of the papermaking process.

cambium   The growing interior of a tree trunk.

chipper   A machine for grinding pieces of wood into small chips.

clearcut   Cutting down all the trees and plants in an area of forest to the bare ground.

contaminants   Non-woody objects like staples or pebbles in a pulp mixture.

cord   A stack of wood that is four feet by four feet by eight feet.

cutting   A small twig cut from a tree to start a new tree.

ecosystem   The interaction of the living and non-living parts of an area like a forest.

fiber   The structural part of wood that makes it strong and flexible.

flume A wooden water-channel used to float logs from one place to another.

graft To attach a cutting from one tree to the limb of another tree.

hoedag A tool for planting trees.

lignin The natural glue in wood that holds wood fibers together.

logger A person who plants and cuts trees for wood and paper companies.

monoculture Growing only one type of tree (or plant) in an area.

overharvesting Cutting too many trees from a forest.

phloem The lining of tree bark.

plywood A building material made by gluing together numerous thin sheets of wood.

pulp The smooth liquid mixture of wood and water that can be pressed into flat sheets to make paper.

ragger The rope that pulls contaminants from a pulping machine.

slurry The liquid made by mixing woods chips with water.

stand A group of trees.

wanigans Bunks that early loggers lived in during a logging job.

For More Information

In The United States

Pulp and Paper Industry
P.O. Box 150559
Grand Rapids, MI 49515
(616) 336-1858
e-mail: directory@pulpandpaper.net
Web site: http://www.pulpandpaper.net

In Canada

Canadian Pulp and Paper Association
1155 Metcalfe Street, 19th Floor
Montreal PQ H3B 4T6
(514) 866-6621
e-mail: communic@cppa.cat
Web site: http://www.open.doors.cppa.ca

Paper Companies

International Paper Company
Two Manhattanville Road
Purchase, NY 10577
(800) 223-1268
e-mail: comm@ipaper.com
Web site: http://www.internationalpaper.com

Environmental Organizations

Rethink Paper
c/o Earth Island Institute
300 Broadway, Suite 28
San Francisco, CA 94133
e-mail: rtp@earthisland.org
Web site: http://www.rethinkpaper.org

Web Sites

Pulp and Paper Online
http://www.pulpandpaperonline.com

Robert C. Williams American Museum
    of Papermaking
http://www.ipst.edu/amp

For Further Reading

Appelbaum, Diana. *Giants in the Land.* Boston: Houghton Mifflin Company, 1993.

Biermann, Christopher J. *Handbook of Pulping and Papermaking.* Orlando, FL: Academic Press, 1996.

Boy Scouts of America. *Pulp and Paper.* Irving, TX: The Boy Scouts of America, 1974.

Dawson, Sophie. *The Art and Craft of Papermaking.* Asheville, NC: Lark Books, 1996.

Drushka, Ken, Hannu Konttinen, and Ken Orushka. *Tracks in the Forest: The Evolution of Logging Equipment.* Claremont, CA: Harbour Publishing Company, 1997.

Durbin, Kathie, *Tree Huggers: Victory, Defeat, and Renewal in the Northwest Ancient Forest Campaign.* Seattle, WA: Mountaineers Books, 1996.

Heilman, Robert Leo. *Overstory: Zero: Real Life in Timber Country.* Seattle, WA: Sasquatch Books, 1995.

Hunter, Dard. *Papermaking: The History and Technique of an Ancient Craft.* Mineola, NY: Dover Publications, 1978.

Meigs, Cornelia. *Swift Rivers.* New York: Walker and Company, 1994.

Stier, Roy E. *Down the Hill: A True Story of Early Logging in the Pacific Northwest.* Wilsonville, OR: BookPartners, 1995.

Toale, Bernard. *The Art of Papermaking.* Worster, MA: Davis Publications, 1983.

Index

About the Author

Allison Stark Draper is a writer and editor. She lives in New York City and the Catskills.

Photo Credits

Cover © Pictor; pp. 2, 18, 31, 34, 36, 48, 52, 54 © SuperStock; pp. 8, 21, 24, 40, 43 © Pictor; p. 11 © Archivo Iconografio, S.A./CORBIS; p. 13 © Bettmann/CORBIS; p. 22 © John Zoiner/Pictor; p. 28 © Lorne Patrick/Photo Researchers; p. 29 © James L. Amos/Photo Researchers; p. 33 © Huntington Library, Art Collections, and Botanical Gardens, San Marino, California/SuperStock; p. 46 © Sally A. Morgan, Ecoscene/CORBIS.

Design

Geri Giordano

www.ingramcontent.com/pod-product-compliance
Lightning Source LLC
Chambersburg PA
CBHW050910210326

41597CB00002B/83